Looking at Clouds

Thick Stratus Clouds

By Kim Thompson

TABLE OF CONTENTS

A Crabtree Seedlings Book

Crabtree Publishing

crabtreebooks.com

School-to-Home Support for Caregivers and Teachers

This book helps children grow by letting them practice reading. Here are a few guiding questions to help the reader build comprehension skills. Possible answers appear here in red.

Before Reading:

- What do I know about clouds?
    - *I know that the sky has different clouds at different times.*
    - *I know that some clouds are storm clouds.*
- What do I want to learn about clouds?
    - *I want to learn about different kinds of clouds.*
    - *I want to know what the weather will be like by looking at clouds.*

During Reading:

- What have I learned so far?
    - *I have learned that clouds are made of water droplets or ice crystals.*
    - *I have learned that stratus clouds can cover the sky like a blanket.*
- I wonder why...
    - *I wonder why clouds form at different heights.*
    - *I wonder why stratus clouds often block the Sun's light.*

After Reading:

- What did I learn about stratus clouds?
    - *I have learned that stratus clouds are the lowest clouds in the sky.*
    - *I have learned that stratus clouds can bring snow or drizzle.*
- Read the book again and look for the glossary words.
    - *I see the word* ***overcast*** *on page 4 and the word* ***drizzle*** *on page 12. The other vocabulary words are found on page 22.*

Thick Stratus Clouds

Clouds are made of tiny droplets of water or ice. One kind of cloud is called stratus.

Stratus clouds form in sheets. They may cover the sky like a thick blanket. The weather is called **overcast** when clouds cover the sky.

Cloud Watch

The word stratus means "layer."

Water heated by the Sun turns into an invisible gas called **water vapor**. The gas rises up high. It cools and turns back into water droplets. They join together to make clouds.

Cloud Watch

Clouds are part of Earth's water cycle. Water moves from oceans, up to the sky to form clouds, then back down to the ground and the oceans.

Stratus clouds usually form low in the sky. They can spread out over a large area at a height that is lower than most airplanes fly.

When a cloud's water droplets grow too heavy to float, they fall. This is called **precipitation**. Rain and snow are kinds of precipitation.

Stratus clouds can bring **drizzle** when it is warm and snow when it is cold.

All clouds move and change. Warm air can change water droplets in a stratus cloud back into a gas. The cloud can disappear before your eyes!

Some clouds higher up also have stratus in their names. Altostratus clouds are gray. This solid layer of clouds makes the Sun look **hazy**.

Cloud Watch

If you see altostratus clouds, lots of precipitation may be falling soon.

Cirrostratus clouds are so high up they are made up of ice crystals. They form in a thin, white sheet. The ice crystals shine like a **halo** around the Sun or the Moon.

Cloud Watch

If you see cirrostratus clouds, a rainstorm or snowstorm may come in the next day or two.

Stratus clouds can even form near the ground! This is called fog. Fog happens when cool air turns water vapor back into water droplets near the ground.

Cloud Watch

It is hard to see in fog because there are so many water droplets in the air. It is like you are standing inside a cloud!

Glossary

drizzle (DRIZ-uhl): Light, misty rain

halo (HAY-loh): A circle or arc of light surrounding an object

hazy (HAY-zee): Unclear or dim

overcast (OH-vur-kast): Covered with clouds; blocking sunlight

precipitation (pri-sip-i-TAY-shuhn): The falling of water from the sky in the form of rain, sleet, hail, or snow

water vapor (WAW-tur VAY-pur): Water that has changed from a liquid into a gas

Pronunciation Guide
altostratus (al-toh-STRAY-tuhs)
cirrostratus (sir-oh-STRAY-tuhs)
stratus (STRAY-tuhs)

Cloud levels		Cloud types
High clouds *Cloud names that begin with "cirro" are the highest clouds in the sky.*	*above 18,000 feet (5,486 meters)*	**High clouds** *cirrus* *cirrostratus* *cirrocumulus*
Middle clouds *Clouds in the middle of the sky have names that start with "alto."*	*6,500 to 18,000 feet (1,981 to 5,486 meters)*	**Middle clouds** *altostratus* *altocumulus*
Low clouds *Clouds that form low in the sky have "strato" in their name.*	*up to 6,500 feet (1,981 meters)*	**Low clouds** *stratus* *stratocumulus* *nimbostratus*

Index

Written by: Kim Thompson
Designed by: Samara Parent
Series Development: James Earley
Proofreader: Kathy Middleton
Educational Consultant: Marie Lemke M.Ed.
Photographs: All images from Shutterstock

Crabtree Publishing

crabtreebooks.com 800-387-7650

Printed in the USA/CP2025

Published in Canada
Crabtree Publishing
616 Welland Avenue
St. Catharines, Ontario
L2M 5V6

Published in the United States
Crabtree Publishing
347 Fifth Avenue
Suite 1402-145
New York, New York, 10016

Library and Archives Canada Cataloguing in Publication
Available at Library and Archives Canada

Library of Congress Cataloging-in-Publication Data
Available at the Library of Congress

Hardcover: 978-1-0398-8124-2
Paperback: 978-1-0398-8485-4
Ebook (pdf): 978-1-0398-8244-7
Epub: 978-1-0398-8364-2
Read-Along: 978-1-0398-8569-1
Audio: 978-1-0398-8527-1